Under the Rain Forest Canopy

Katie Peters

GRL Consultants,
Diane Craig and Monica Marx,
Certified Literacy Specialists

Lerner Publications ◆ Minneapolis

Note from a GRL Consultant
This Pull Ahead leveled book has been carefully designed for beginning readers. A team of guided reading literacy experts has reviewed and leveled the book to ensure readers pull ahead and experience success.

Copyright © 2020 by Lerner Publishing Group, Inc.

All rights reserved. International copyright secured. No part of this book may be reproduced, stored in a retrieval system, or transmitted in any form or by any means—electronic, mechanical, photocopying, recording, or otherwise—without the prior written permission of Lerner Publishing Group, Inc., except for the inclusion of brief quotations in an acknowledged review.

Lerner Publications Company
A division of Lerner Publishing Group, Inc.
241 First Avenue North
Minneapolis, MN 55401 USA

For reading levels and more information, look up this title at www.lernerbooks.com.

Main body text set in Memphis Pro 24/39
Typeface provided by Linotype.

Photo Acknowledgments
The images in this book are used with the permission of: © iStockphoto (all)

Front cover: © Shutterstock

Library of Congress Cataloging-in-Publication Data

Names: Peters, Katie, author.
Title: Under the rain forest canopy / Katie Peters.
Description: Minneapolis : Lerner Publications, [2020] | Series: Let's look at animal habitats (Pull ahead readers - Nonfiction) | Includes index. | Audience: Age 4–7. | Audience: K to Grade 3.
Identifiers: LCCN 2018059816 (print) | LCCN 2019000024 (ebook) | ISBN 9781541562059 (eb pdf) | ISBN 9781541558595 (lb : alk. paper) | ISBN 9781541573147 (pb : alk. paper)
Subjects: LCSH: Rain forest animals—Juvenile literature.
Classification: LCC QL112 (ebook) | LCC QL112 .P475 2020 (print) | DDC 591.734—dc23

LC record available at https://lccn.loc.gov/2018059816

Manufactured in the United States of America
1 – CG – 7/15/19

Contents

Under the Rain Forest
  Canopy 4

Did You See It? 16

Index 16

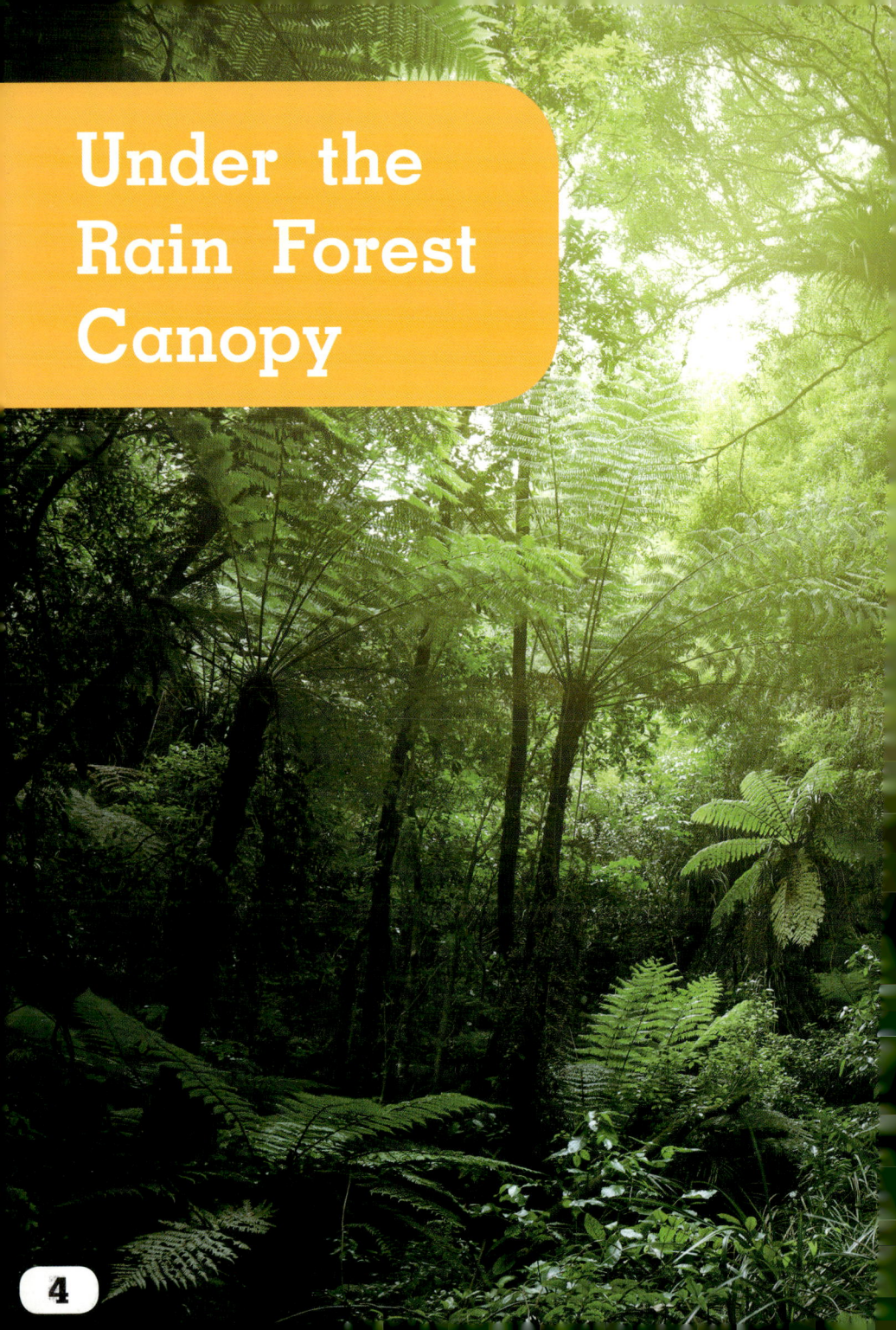

Under the Rain Forest Canopy

The rain forest has tall trees. The leaves make a canopy.

Monkeys play in the trees.

Sloths climb in the trees.
The sloths move slowly.

Parrots rest in the trees.

Frogs sit in the trees.

Jaguars hide in the trees.
Do you see the jaguar?

Did You See It?

frog

jaguar

monkey

parrot

sloth

tree

Index

canopy, 5
frogs, 13
jaguars, 15
leaves, 5

monkeys, 7
parrots, 11
sloths, 9